Clam Bridges of Dartmoor

Peter F. Mason

Additional text by Chris Walpole

Cover image: Clam bridge at *Hawns and Dendles, Ivybridge* (detail)
© Paul Rendell Collection

Published by the Lustleigh Society
www.lustleigh-society.org.uk
2019

ISBN 978-0-9957122-4-9

Designed and printed by Short Run Press Ltd,
Bittern Road, Sowton, Exeter, EX2 7LN

The writing and publishing of this book was a Lustleigh *Parishscapes* project, funded through the *Moor than meets the eye* scheme, supported by the National Lottery Heritage Fund. Through *Moor than meets the eye* people have explored Dartmoor's rich past, worked to conserve its wildlife and archaeology, improved understanding of the landscape and developed and shared the skills needed to look after it for generations to come. Active over a 280 square km area, *Moor than meets the eye* has told 'The Dartmoor Story' of the last 4000 years.

Emma Stockley
Moor than meets the eye Community Heritage Officer
August 2019

Foreword

Although there was some debate in the nineteenth century about what was meant by a 'clam' bridge, in the Dartmoor region the term undoubtedly referred to a wooden bridge. Often they were modest trunk or plank bridges with a single handrail, but all enabled people to cross rivers and streams safely, even when in spate, and were vital assets for many communities, connecting homes with churches, farms, mills, mines and markets.

The great medieval stone bridges spanning the River Dart probably had timber precursors, and the tradition of constructing wooden bridges survived in many parishes, including locations well-frequented by tourists from around 1800 onwards, such as at Becky Falls and Lydford Gorge. Most clam bridges were small in scale, and often picturesque.

As a resident of Lustleigh, Peter Mason joined a group who managed to save the historic clam bridge across the River Bovey in Lustleigh Cleave in 2006–7, when it was threatened with removal and replacement. In the event a massive and expensive new bridge was built beside it, but the clam bridge was retained. As early as 1904 the parish council of Lustleigh had recognised the historical importance of this bridge as probably the last survivor in the Dartmoor region of a traditional footbridge built from tree-trunks and supplied with a handrail.

The successful campaign inspired Peter Mason to research other examples of these wooden bridges in the Dartmoor region. Drawing on artists' depictions from the 18th century onwards, and photographs from the mid-19th century to the present day, as well as contemporary descriptions and fieldnames, he has brought together, for the first time, a very welcome and richly illustrated account of these significant bridges. A comprehensive study (by Chris Walpole) of bridges within Belstone parish is included, and implies, excitingly, that further research will reveal more sites elsewhere.

Tom Greeves, Tavistock, June 2019

Introduction

Historically, the term 'clam' has been used of both stone and wooden bridges in Britain, and this still applies today. For example, if the words 'clam bridge' are typed into the Google search engine, the first place to be listed is Wycoller, a village in Lancashire. The entry reads:

> Various ancient bridges cross Wycoller Beck, including 'Pack-Horse Bridge', a twin arched bridge in the centre of the village, 'Clapper Bridge' and 'Clam Bridge'.[1]

1 *The clapper bridge at Postbridge*
(© Dartmoor National Park Authority)

The bridge illustrated as the 'Clam Bridge' is made of stone and is similar to the clapper bridges on Dartmoor. Visitors to Dartmoor will be familiar with the most famous of these at Postbridge (Fig. 1). Sometimes shown on maps or described in books as being "Cyclopean"[2] this is typical of the clapper bridges of Dartmoor and, as we've seen by the entry for Wycoller, this type of bridge is not exclusive to Dartmoor.

However, although there has been debate over the origin and meaning of the word 'clam' in the context of bridges, here in the West Country at least, it has most commonly been applied to wooden bridges and this nineteenth century image shows the traditional construction (Fig. 2). An early and highly improbable depiction of a clam bridge, possibly over the River Bovey in Lustleigh Cleave, was published in 1822 (Fig. 3).

The *Oxford English Dictionary*[3] defines a clam bridge as "a plank or crossing-stone over a brook." The word being "short for clammer or clamber, used of a footbridge". The dictionary gives as examples of its use extracts from *Exmoor Scolding* published in 1746: 'Dest'nt remember when tha com'st over the clam… whan tha Wawter wos by stave."; and from S. Smiles *Lives* [of the] *Engineers* published in 1861: "There is

2 *Rustic Bridge near Tavistock (Adeney)* Ward & Lock's Pictorial and Historical Guide to Dartmoor *1888* (© Devon Archives & Local Studies Service SB/DAR5/1888/WAR)

3 *Clam Bridge* (Lithograph) Thomas Hewitt Williams, Published by Cole & Co., Exeter 1822 (© Devon Archives & Local Studies Service P&D09197)

a fourth [bridge] on the Blackabrook[4] consisting of a single stone or clam."

Joseph Wright's *English Dialect Dictionary* of 1898 defined a clam as: "A bridge formed of a plank or the trunk of a tree" <u>or</u>: "A stone slab laid across a stream, a stepping stone."[5] Wright goes on to give quotations to support each definition: one for each from *An Exploration of Dartmoor* by John Warden Page[6]. He also gives a definition of a 'clammer' as being: "A pole or plank laid across a stream as a footbridge." Although

this was in use in some counties in the north of England he quotes an example of its use by a local person giving directions in west Somerset in 1883: "You'll come to a clammer, and tother zide o' the river the path's plain enough." Clem Marten in his dictionary of Devon dialect includes 'clammer' in a glossary describing it as "a footbridge, examples made of stone are found on Dartmoor"[7].

As is apparent from these sources the use of the word 'clam' is mainly limited to the West Country. An early local description of a wooden bridge comes from Mrs Bray's guide to the Tamar and Tavy of 1836 in which she poetically describes the bridge over the River Walkham, just above the confluence with the River Tavy at Double Waters:

There is a miniature alpine bridge that crosses the Walkham at its junction with the Tavy … this consists of a single plank with a light piece of wood extended as a hand rail to hold by in passing. In one part the plank is supported by a *clutter* of rocks beneath, as a Devonian would say in describing it. To stop on the middle of this plank and look around will afford the greatest delight to the lover of the picturesque … These bridges are called *clams*, and they are never found anywhere excepting across our rocky and mountain streams."[8]

Fig. 4 shows the bridge towards the end of the nineteenth or early in the twentieth century and Fig. 5 shows it in 1962 (with a young Tom Greeves!).

4   *Virtuous Lady Mine, River Walkham* Welch & Sons, Portsmouth, 1907 or earlier (© Greeves Collection)

5   *Bridge at Double Waters, 1962* (© Greeves Collection)

At the end of the nineteenth century there was a lively debate over the meaning of the word 'clam' in the context of bridges. In the 1880s William Crossing was engaged in correspondence about the subject in the *Western Antiquary*. He wrote:

> In this part of Devonshire, the word [clam] is sometimes applied, both to the old stone bridges of Dartmoor, and to light wooden foot bridges, but the former are oftener known by the name of *clapper* bridges, a term which I never knew applied to the wooden ones, always having heard them spoken of as *clams*.[9]

Another correspondent to the magazine, Charles Harris, took issue with this, quoting Samuel Rowe as writing in *A Perambulation of Dartmoor*:

> Some of these [bridges] are formed of a singular stone, and would then probably come under the vernacular denomination, *clam*; a term also frequently applied to a bridge formed of a plank, or single tree, although I have noticed a distinction sometimes made, the *wooden* bridge being called a Clapper, and the *stone* bridge, a Clam.[10]

In April 1882, Crossing replies at length, stressing how familiar he is with Dartmoor and how much time he has spent there and, most importantly, the amount of time he has spent talking to people living on Dartmoor:

> I have passed many an evening by the peat-piled hearth, and heard numberless tales of the moor-lands, from the dwellers in those lonely hill-farms, so I have certainly had *some* opportunity of hearing the words *clam* and *clapper* applied. At the same time, I should not like to say, because I have never heard it, that a wooden bridge was never called a *clapper*, but if it is sometimes so, it is by no means the general term for it.[11]

Crossing goes on to quote Worth and Dymond as speaking of the stone bridges as 'clappers' and refers to a wooden bridge over the Avon near Diptford being called *The Clam Bridge*. He also quotes Sir John MacLean having described a *clam* as 'a stick laid across a stream of water.' However, in the same edition FB Doveton remembers hearing "Perritt [sic] (the Dartmoor guide and a canny fellow!) calling a small *wooden* bridge just below North Bovey 'the Clap Bridge' and always speaking of the Cyclopean stone ones as 'clams'."

In his book about Dartmoor Hansford Worth describes a clam bridge as being "for foot passengers"[12] and Crossing in his *Guide to Dartmoor* defines a clam as "a wooden footbridge, seldom seen on Dartmoor"[13].

The purpose of a clam bridge is basically no different to that of a clapper bridge and there are, I believe, two reasons for Crossing's remark about them being seldom seen on Dartmoor – firstly that they were probably quite common in the rest of Devon and on the fringes of the moor where timber was readily available and stone less so and secondly, that the opposite applied on the open moor.

Having considered all aspects of this debate I have taken Crossing's view on the meaning of the word 'clam' to be definitive. Therefore the definition that I have used is that clam bridges are wooden footbridges, some more elaborate than others and sometimes of a rustic design aimed to appeal to a feeling of the 'Picturesque', but always made of wood. This study looks mainly at historic examples of these bridges, particularly where there is photographic evidence.

7 *Otterton Park, March 1795* Reverend John Swete
(© Devon Archives & Local Studies Service DRO564M/F8/17)

6 *Near Chulmleigh, Devon* 1862, John Wallace Tucker
(© Royal Albert Memorial Museum and Art Gallery, Exeter)

Away from the moor elsewhere in Devon, there are examples of clam bridges both from maps, court records and images, including the following examples. One of the longest and possibly most extraordinary was captured in a painting of 1862 by John Wallace Tucker (Fig. 6).

Another clam bridge appears in a painting of Otterton Park by the Reverend John Swete of 1795 (Fig. 7). The bridge of wooden construction crosses the River Otter adjacent to a plot named 'Clamour' on the Tithe Map and in a report by the Admiralty of 1851 concerning a proposal for an embankment it marks and names it as 'clamor' bridge.[14]

Fig. 8 shows a typical clam bridge from an unknown location, possibly in north Somerset. The photograph was taken by Thomas Bunce of Bristol (1865–1928) and shows his son, Herbert, on the bridge. This is a classic simple clam bridge of one substantial piece of wood laid across the stream with a single handrail.

8 *An unknown clam, c1901–1905*
(Courtesy Sue Wilson née Bunce ©)

In 1748 John Matthews of Abbotskerswell, a yeoman, was summoned to the Quarter Sessions for not having repaired "a common footbridge made of timber and commonly called a clam situated upon a river of water in the King's common highway leading from Abbotskerswell to Kingskerswell and for which he is held responsible by reasons of his tenure of lands adjoining to and surrounding the footbridge."[15]

Clam bridges on Dartmoor

The majority of the evidence for historic clams on Dartmoor is provided by photographs and postcards dating from the second half of the nineteenth century and the beginning of the twentieth. Field names can also indicate the possible former existence of a clam bridge and we will look at this in the next section. As far as I have been able to ascertain there is only one traditionally built clam bridge (i.e. made of tree trunks laid across the river with a single handrail) remaining on or near to a public right of way and this is discussed in the final section.

Detailed examination of the old 6" or 25" inch Ordnance Survey maps of Dartmoor, mostly surveyed in the late 19th century can also reveal the sites of footbridges. The majority of these, at least on the edge of the moor, were most likely to have been made of wood, i.e. they were clam bridges. This is particularly evident from the detailed research that has been carried out by Chris Walpole in Belstone parish and the valleys of the River Taw and East Okement River. This research is included in full as a case study.

The research for this publication indicates that there

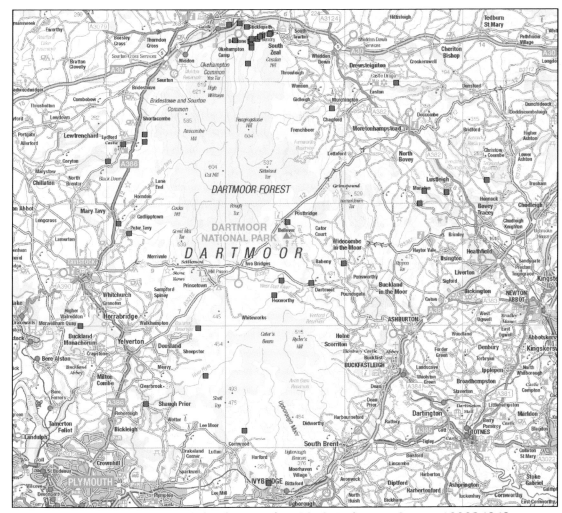

9 *Map showing locations of clam bridges on Dartmoor (Courtesy of the Dartmoor National Park Authority)*

Crown copyright and database rights 2019 Ordnance Survey 100024842

were, at the very least, 37* locations where there have been clam bridges in the area defined by the Dartmoor National Park boundary. As can be seen from the map (Fig. 9) these were mainly to be found on the fringes of the moor or in wooded valleys. This supports the view that these footbridges were constructed from the most easily available material in the area.

One of the earliest images is a watercolour painting of a bridge near Ivybridge by an unknown artist[†] (Fig. 10). Again it shows the classic traditional construction of a clam bridge, with tree trunks forming the main

11   *In the woods, Ivybridge*
(© Ivybridge Heritage and Archive Group)

construction. The bridge is likely to have spanned the River Erme north of Ivybridge at approximately SX 636 571. Fig. 11 is possibly a photograph of a bridge at a location in the same area in the twentieth century.

Becky Falls, Becka Brook

One of the most photographed clam bridges was located at Becky Falls, near Manaton (Fig. 12) (SX 763 800). It is not surprising that it was photographed a lot as the site was a much visited tourist spot on the fringes of the moor in both the nineteenth and twentieth

10   *View of the Glen near Ivybridge 1780*
(© The British Library Board (Maps K.Top 11.118m)

*Although these were not necessarily all in existence at the same time this is probably a significant underestimate. Also note that all grid references are approximate.
†The British Library catalogue gives the initials of the artist as J.F.E.

12 *Becka Falls* by Francis Bedford c. 1865
(© Greeves Collection)

centuries. It first became popular in the nineteenth century when visitors sought out places that matched the criteria for the 'Picturesque'. By the time Samuel Rowe, vicar of Crediton, visited Lustleigh on his way from Moreton to Bovey, the Cleave was already, in his words, 'far-famed'[16] and Becky Falls was one of the sites to be seen when visiting the area.

The first through train reached Exeter on 1st May 1844 and the railway network was extended to Dartmoor in the coming years. The first edition of Murray's *Handbook for Travellers in Devon and Cornwall* was published in 1851 and the guide had run to six editions by 1865. The author of the guide described the spot as being "at all times romantic and delightful …" and that the Falls were to be "approached over a rustic bridge by a field path from the lane …".[17]

From a close look at the surroundings it appears that the postcard (Fig. 13) may have been reproduced from a photograph by Francis Bedford taken on the same day as the image in Fig. 12.

Near Dartmoor.   July 1907.   *Becky Fall Rustic Bridge.*

The Wrench Series, No. 2109

13 *Becky Fall Rustic Bridge*, postmarked 1907
(© Paul Rendell Collection)

14 *Becky Bridge*, Francis Frith postcard, 1907 or earlier
(© Greeves Collection)

15 *Becky Falls, Dartmoor*, unknown date (© Paul Rendell Collection)

16 *Becky Bridge*, unknown date (© The Dartmoor Archive)

17 *Becky Bridge*, unknown date (© The Dartmoor Archive)

Lydford, River Lyd

On the other side of the moor, Lydford Gorge, the deepest in the South West of England, was also a site that attracted visitors from the earliest days of tourism in Devon. In 1889 Page, the author of *An Exploration of Dartmoor and its Antiquities* wrote:

> Than this romantic gorge no fairer spot exists in the West Country, and few more alarming to a head other than strong. To within a few feet of the boiling, eddying stream the precipice is clothed with ferns and underwood, the delicate tracery of branches far

18 *Lydford Gorge*, photo by Furze, c.1860 (© Greeves Collection)

up against the sky almost concealing the bridge from view. The rocks are black with perpetual moisture, and slippery as glass under the ceaseless friction to which they are subjected by the angry river. Notwithstanding occasional protection in the shape of a handrail, the timid are recommended not to approach the bottom of the cleft, as a false step might very well prove fatal.[18]

19 *The Waterfall, Lydford* Chapman postcard 3641
(© The National Trust)

20 *Lydford Gorge Rustic Bridge* Frith postcard
(© The National Trust)

Described in Murray's Handbook as a "chasm" visitors were encouraged "to scramble as far as possible down the rocks … to obtain a good view of the singular scene … and to ascend the course of the river to *Kitt's Fall*."[19] A very attractive feature for those in search of the 'Picturesque' and the 'Sublime' were the White Lady Falls, although one of the earliest travellers to write about them, William Gilpin, the writer responsible for defining the 'Picturesque' in the eighteenth century, was underwhelmed:

> The *fall of the river* which brought us hither, and which is the least considerable part of the scenery … is a mere garden-scene. The steep woody hill, whose shaggy sides we had descended, forms at the bottom, in one of its envelopes, a sort of little woody theatre; rather indeed too lofty when compared with its breadth, if Nature had been as exact as Art would have been, in observing proportion. Down the central part of it, which is lined with smooth rock, the river falls. This rocky cheek is narrow at the top, but widens as it descends, taking probably the form of the stream when it is full. At the time we saw it, it was rather a spout than a cascade; for though it slides down a hundred and eighty feet,[‡] it does not meet one obstruction in its whole course, except a little cheek in the middle.[20]

However, Gilpin was probably in a minority. For example, Samuel Rowe, who visited Lydford in the 1820s, described the waterfall as being "celebrated".[21] Fig. 19 shows White Lady Falls with the bridge in the foreground and Fig. 20 shows the same bridge from the approach to it looking upstream.

Figs. 21 and 22 show the same bridge from the same viewpoints in what was possibly a later simpler form.

There is no longer a bridge at the location of the bridges shown in those images (Figs. 19–22). The bridge

21 *The Waterfall, Lydford Gorge, Dartmoor* Fac-Sim Oil-Colour postcard from a watercolour by Herbert Truman, EA Sweetman and Sons Ltd (© The National Trust)

‡In fact White Lady Falls are only approximately 100 feet in height.

22 *Lydford footbridge near Cascade* Frith postcard (© The National Trust)

23 British Mirror Series postcard (© The National Trust)

shown in Fig. 23 possibly crossed the River Lyd itself near White Lady Falls (SX 501 835).

Figs. 24 and 25 are postcards of a bridge towards the southern end of the Gorge (SX 514 846). Fig. 24 is looking upstream towards the Devil's Cauldron and Fig. 25 is looking downstream at Pixie's Glen.

24 *The Lydford Gorge, Devon* (© Paul Rendell Collection)

Also portrayed near Lydford was a clam bridge at Lydford Mill (SX 514 849). Mills were particularly attractive to artists as their rustic nature appealed to the lovers of the 'Picturesque'. The first postcard (Fig. 26), taken from a tinted photograph, was probably published in the early 1900s.

25 *Lydford Gorge, the pool below the Gorge* (© Paul Rendell Collection)

26 *Lydford Mill* date unknown (© Greeves Collection)

27   *Lydford Mill*, postcard pre 1907
(© Paul Rendell Collection)

28   *R. Lyd*, Lydford Chapman postcard 12715
(© The National Trust)

The second postcard (Fig. 27) probably shows a different bridge. This image is taken from a painting by Ernest William Haslehust (1866–1949). Haslehust was a prolific artist in watercolours who was best known for illustrating the travel book series *Beautiful England*, published by Blackie & Sons, as well as designing posters for railway companies. The composition with a woman crossing the bridge with basket under her arm is a classic one. The picture also shows the simple construction of many clam bridges.

Fig. 28 is an image of another bridge over the River Lyd at a location that has yet to be identified.

Holy Street Mill, Chagford, River Teign

Holy Street Mill, outside Chagford, was another 'Picturesque' scene popular with both artists and photographers in the nineteenth century – to the extent that it was once described as being 'infested with artists'[22]. A footbridge across the River Teign led to the mill as seen in these early photographs by Way and Sons and William Spreat (Figs. 29 & 30) (SX 689 878). In a photograph taken a few years later by W. Widger the handrails are damaged or absent (Fig. 31).

29 *Holy Street Bridge*, Way & Sons, c1860
(© Greeves Collection)

31 *Holy Street Bridge*, W. Widger c.1870 (© Greeves Collection)

30 *Holy Street Bridge*
William Spreat, c.1865
(© Greeves Collection)

Hawns and Dendles, River Yealm

Situated near Cornwood on the south western edge of Dartmoor, Dendles Wood is a Site of Special Scientific Interest. Dendles Wood and the adjacent Hawns Wood are sometimes known collectively as Hawns and Dendles. In *High Dartmoor* Eric Hemery described "The glen of Hawns and Dendles [as being] unique. To describe it as a lost fairyland may give a hint as to its character. ... Fallen trees bridge cascades and tree-shadowed pools; the east bank becomes a fern-festooned cliff, and a deserted ford links the west bank with the green path leading up the valley from Coombe."[23]

On postcards of the clam bridge the site is sometimes referred to as Awns and Dendles as in Fig. 33 (SX 616 614). Fig 32 shows the same bridge at a different time. Although there is more greenery suggesting that it is a later image, the steps up to the bridge are wooden, whereas in the Fig. 33 they are made of stone, indicating that Fig. 33 shows the bridge at a later date. The photograph of a clam bridge at the Upper Falls in Hawns and Dendles (Fig. 34) is much earlier.

32   *Hawns and Dendles, Ivybridge* (© Paul Rendell Collection)

33 *Awns and Dendles, Cornwood* (© Paul Rendell Collection)

34 *Hawns and Dendles, Upper Falls.* Stereoview by RP Yeo c. 1865 (detail) (© Greeves Collection)

Dartmeet, River Dart

35 *Clam Bridge, Near Dartmeet* (© Greeves Collection)

The clam bridge near Dartmeet (Fig. 35), photographed possibly in the 1920s or 1930s, is at a spot that hasn't, at the time of writing, been located. It is probably on the East Dart, possibly at a point near Brimpts Farm or adjacent to Badger's Holt (possibly around SX 672 735). Evidently there are iron fixings on the river bank approximately 200 yards upstream from Badger's Holt which may indicate its location. There are no footbridges in the area marked on the 6" Ordnance Survey maps surveyed at the turn of the nineteenth century.

The Swincombe 'Fairy Bridge', River Swincombe

36   *The bridge over the River Swincombe in its present form.*
(Photo the author)

This location has always been a preferred crossing point of the river Swincombe (SX 641 725). It lies on the route of the medieval Tavistock to Ashburton packhorse track (Fig. 36). In later years miners from Mary Tavy would cross at this point to get to the Hensroost and Hooten Wheals mines. At first the river could only be crossed by the ford or by stepping stones. Whether the river was passable or not depended on the height of the water so, in order to ensure safe passage, a wooden bridge was constructed in 1892. The timber for the

bridge came from the Sherberton Plantation.[24] William Crossing wrote about the construction of the bridge in *Dartmoor Worker*:

> From Princetown the Mary Tavy men followed the route across Tor Royal newtake to Swincombe, where there is a ford over the stream of that name, and also stepping-stones. But during a heavy flood it was not possible to cross by means of the latter and a wooden footbridge was therefore erected.[25]

Although there would probably have been replacements in the interim, the next record we have is when a new bridge was erected at the site in 1972 (Figs. 37 &

37   *Swincombe Stepping Stones and Clam Bridge*
(© Courtesy PH-L Collection)

38). This version was constructed from telegraph poles, planks and a wire handrail. Sometime shortly after it was built Eric Hemery commented that the "Fairy Bridge trembles unnervingly when trodden by several persons simultaneously ...".[26]

This bridge lasted, in one form or another, until 1995 when a more sturdy replacement was installed. This version commemorated the 30th anniversary of

38 *Swincombe Clam Bridge* (© Courtesy PH-L Collection)

the Dartmoor Expedition Panel. To mark this a plaque was placed on a post at the end of the bridge with the inscription: SWINCOMBE BRIDGE WAS REPLACED ON 23rd OF AUGUST 1995 TO COMMEMORATE THE 30th ANNIVERSARY OF THE DARTMOOR EXPEDITION PANEL 1965–1995.[27] In 2014 the bridge was washed away in the

February storms and replaced with the current version (Fig. 36).

The earliest reference to it being called the Fairy Bridge is by Crossing in the early 1900s, though he didn't give an explanation.[28] However, there are ancient beliefs in water deities on Dartmoor and Crossing writes about these in his book on the folklore and legends of Dartmoor:

> The spirit of the Dart ... is said to claim a heart every year ... This was once a popular belief, and there are many who still think it true ... When the 'cry' of Dart – the article is often omitted by the moor people – rises from the valley, it is the spirit of the river calling for his prey ... All the Dartmoor rivers have their 'cry', and this is heard chiefly in the quiet of the evening.[29]

It is therefore not too far-fetched to assume that someone decided that there was a fairy who provided protection to those crossing the bridge. Tim Sandles on his Legendary Dartmoor website describes the appearance of fairy figures under the bridge:

> ... sometime in the early 2000s a small figure of a fairy appeared under the bridge. For a good eight years she peacefully sat under the bridge providing protection for those crossing the river. Then in 2008 she got damaged, whether as a result of trying to withstand the Dartmoor weather and the river Swincombe's

angry spates or at the hand of a human nobody knows. Luckily the following year a replacement was found and installed under the bridge, once again keeping a watch on all moorland travellers. Sadly she also suffered the same fate as her previous guardian and so yet another fairy was found and installed but once again she too did not last long.[30]

West Dart

On the West Dart, just above the point where the Swincombe joins it, there are the remains of a footbridge at SX 647 737 (Fig. 39). A photograph (Fig. 40) taken of it in the early 1900s shows a view looking upstream towards Brownberry Farm which no longer exists. In the late nineteenth and early twentieth centuries the bridge was used by tin miners travelling between Postbridge and Hexworthy Mine.[31]

The bridge was on a footpath running from Sherberton to Dunnabridge that is marked on the 1906 six-inch Ordnance Survey map. Crossing refers to this clam as being a means of crossing the Dart when walking from Brownberry to Sherberton[32] as can be seen on the map (Fig. 41).

Crossing also wrote about it in *Amid Devonia's Alps*:

At a short distance above the confluence of the Swincombe River with the Dart … there is now a wooden foot bridge or clam, thrown over the latter

39 *The remains of the bridge on the West Dart* (Photo: the author)

40 *Footbridge near Sherberton on the West Dart* (©Stanbrook Collection)

41 *Ordnance Survey 6" map Devonshire CVII.NE 1906* (extract reproduced by permission of the National Library of Scotland ©)

42 *Footbridge near Sherberton on the West Dart* (© Ivybridge Heritage and Archive Group)

43 *Footbridge near Sherberton on the West Dart* (© Ivybridge Heritage and Archive Group)

stream, but this did not exist at the time of which I am writing[§], having been only erected a few years.[33]

This suggests that the bridge was built sometime between 1874 and 1888.

The bridge was still standing in the 1950s as seen in the photographs from that period (Figs. 42 & 43).

[§]Crossing is describing a walk he undertook with a friend in 1874 (see chapter 4 of *Amid Devonias's Alps*).

Mary Tavy/Peter Tavy, River Tavy

44 *Mary Tavy Clam* (© Paul Rendell Collection)

45 *Mary Tavy Clam* (© Paul Rendell Collection)

The construction of the bridge near Mary Tavy at SX 510 784 (Figs. 44 & 45) is similar to the method used today to construct footbridges on Dartmoor but, as can be seen on the captions it was regarded as being a clam bridge. The photograph in Fig.45, which was taken from a different viewpoint, is an earlier image as it shows the buildings of Devon United Mine in the background.

There is also an earlier romantic image from 1846 of a bridge in the same location (Fig. 46) which Rachel

46 *Clam Petertavy*
(© Devon Archives & Local Studies Service SC1890)

Evans wrote about in 1846: '... we reached the long rustic bridge or "clam" which crosses the stream [of the R. Tavy] ...'[34]

Also on the River Tavy but at an unknown location is an early photograph by Furze, about whom little or nothing is known (Fig. 47).

Peter Tavy Combe, Colly Brook

48 *Bridge over the Colly Brook near Peter Tavy* (© John Walling Collection)

47 *On the Tavy Photo* by Furze c. 1860 (© Greeves Collection)

49 *Dartmoor, Peter Tavy Coombe* (© Paul Rendell Collection)

Fingle Bridge tea shelter, c. 1912 (© Stanbrook Collection)

The clam bridge over the Colly Brook near Peter Tavy at SX 519 776 (Fig.48) has had different forms over the years as can be seen in these images. Fig. 49 shows it in an earlier, simpler form.

Fingle Bridge Inn, River Teign

Early photographs of the Inn at Fingle Bridge near Drewsteignton show a clam bridge crossing the little stream that runs into the Teign at the Bridge (Fig. 50) (SX 743 899).

Having sold pots of tea in the open air for ten years Jessie Ashplant of Drewsteignton built the first tea shelter in 1907. Evidently, because the landowner thought the roof of corrugated iron was unsightly, Jessie covered it with furze and bracken. That shelter lasted until 1929 and its replacement lasted until 1957 when the first of the inns on the site was built.[35]

Shaugh Prior and the River Plym

Crossing describes the area around Shaugh Bridge (SX 533 636), the spot from which Fig. 51 was photographed, as being "long celebrated as one of the beauty spots of the moorland borders."[36] Elsewhere he states that: "Above the confluence [with the River Meavy] the Plym is spanned by a clam, and from this a path leads up to the summit of the hill which forms the southern extremity of Wigford Down …".[37]

51 *The River Plym in spate from Shaugh Bridge, April 1973*
(© PH-L Collection)

52 *Shaugh Bridge, Photocrom Ltd* (© Greeves Collection)

Shaugh Bridge which is referred to in the caption for Fig. 52 can be seen in the background between the trees. From a close examination it would appear that the view is looking down the River Meavy rather than the Plym and this suggests that it is a picture of a clam bridge spanning the Meavy.

Murray, in his Handbook for Travellers, recommends Shaugh Bridge Station on the line from Plymouth to Tavistock as a stopping off point to reach the Plym and the Dewerstone: "This is a singularly wild and romantic spot, where the *Mew*** and *Cad* unite their noisy streams among antique oaks and rocks and take the name of Plym. It highly deserves the attention of artists"[38]. If Fig. 52 is indeed a bridge over the River Meavy it would have acted as a short cut to the Dewerstone and Dewerstone Cottage which, apparently, used to serve cream teas.

Whilst discussing bridges over the River Meavy it is worth including the mention by Worth of a clam for 'foot passengers' in Meavy Parish, "a little over a mile below Marchant's Bridge, [which] afforded the only crossing [of the River Meavy at this point] until, in 1887, a substantial arched bridge was constructed".[39] This bridge, the Lower Meavy Bridge, was replaced by an iron construction in 1909.[40]

**An alternative name for the River Meavy.

Foxworthy, River Bovey

This photograph (Fig. 53) was taken by a member of the Hunt family, who lived at Foxworthy in the Bovey Valley. Both A. R. Hunt and his son, Cecil Arthur Hunt who was an artist, were photographers. The exact location of the bridge is not known. The Ordnance Survey 6" map surveyed in 1885 shows a footbridge, which is mentioned by Crossing in his *Guide to Dartmoor*[41], over the River Bovey approximately half way between Horsham Steps and the 'Clam Bridge'. However the river is quite wide at this point – certainly wider than the stream being crossed in the photograph. It may therefore have been over a small tributary nearer Foxworthy itself.

River Wrey, Lustleigh

This bridge (Fig. 54) was likely to have been over the River Wrey on the footpath that runs from Wreyland to Mill Lane shown on the Ordnance Survey map (Fig. 55) (SX 786 809). The Chapman postcard dates from the early years of the twentieth century. The use of the word 'cleal' on the caption is interesting. It doesn't appear in the Oxford English Dictionary or in Partridge's *Dictionary of Slang and Unconventional English*. Therefore the question arises as to whether it was a name local to the people of Lustleigh or a misprint by the publisher of the postcard.

53 *River Bovey, near Foxworthy* (© Simon Butler Collection)

Cleal Bridge, Lustleigh (© Lustleigh Community Archive)

55 *Ordnance Survey 25" map Devonshire C8.NE 1885* (extract reproduced by permission of the National Library of Scotland ©)

Kingsett Clam, Newleycombe Lake

A footbridge is marked on the Ordnance Survey Map of 1886 (SX 577 698) to the south east of Kingsett Farm near Burrator spanning the Newleycombe Lake (Fig. 56). As is often the case, the clam crosses the river just above a ford – now largely obliterated. Mike Brown gave details of it in his *Dartmoor Field Guides*:

Just above the ford two granite blocks narrow the stream's channel considerably, and one still bears an iron ring – the other has a broken stump – on the upper face indicating the site of an earlier clam, Kingsett

56 *Ordnance Survey 6" Map CX11.NE 1886 (extract reproduced by permission of the National Library of Scotland ©)*

Bridge. This must have been a very short one, for the stream can be very easily stepped across here, and the gap could in fact have been more readily spanned by a single clapper. The rings were the mountings for chains fixed to hold the clam secure against spates, the lack of a bridge here indicating that they were not wholly successful in achieving the desired effect![42]

This bridge would have needed regular replacement and the daughter of the Pearse family, who lived at Kingsett remembers a wooden plank being put there in the early 1920s.[43] The photograph (Fig. 57) shows the spot today.

57 *Site of Kingsett Bridge* (photo Paul Rendell)

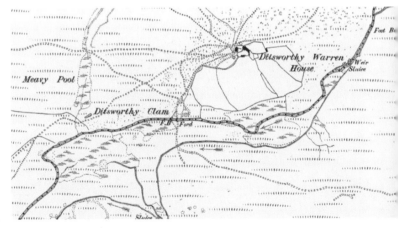

58 *Ordnance Survey 6" Map, Devonshire CXII.SE 1886* (extract reproduced by permission of the National Library of Scotland ©)

Ditsworthy Warren, River Plym

In writing about an excursion that took in Ditsworthy Warren House, Crossing describes the need to cross the River Plym by the Ditsworthy Clam shown on the Ordnance Survey map (Fig. 58) (SX 583 659). Bearing in mind the fact that Crossing was very particular in his definitions and that, in the description of the same excursion, he had described a bridge over the Blackabrook as "a single stone clapper" it is more than likely that the Ditsworthy Clam was of timber construction.[44]

Belstone parish and the valleys of the River Taw and East Okement River

Over the years there have been a number of clam bridges, almost certainly all wooden or mostly wooden, in the Belstone area. There is evidence of fourteen footbridges on the River Taw between the Belstone/ Sticklepath parish boundary and Taw Marsh – only three of which have modern equivalents. There have been two footbridges – both with modern equivalents – over the East Okement River where it forms the western boundary of Belstone parish, and a further three – none of which still exist – within the parish on the small un-named tributary which flows from above Priestacott hamlet down to the East Okement at Fatherford.

Working upstream from Sticklepath there were two temporary bridges for which there is no photographic evidence. The first of these was remembered by John Bowden, who has lived beside the Taw at *Skaigh View* for many years:

> Once during the Second World War lots of the ash and oak trees in Skaigh Wood were removed 'for the war effort'. A temporary bridge was erected just up from *Skaigh View* which was strong enough to take the lorries that took the trunks away. The driver of the tractor that extracted the trees used to give the children rides. The tops of the cut trees were sold to locals as firewood.

This bridge would have been at SX 638 940. The second bridge, which was situated in approximately the same place, was built by John Bowden across the Taw by *Skaigh View.* This lasted for several years until washed away in a storm in 1952. When Sticklepath parishioners carried out their Beating of the Bounds walk for the first time on 9 September 2000 (and again on 19 August 2007 and 17 August 2014) they crossed the Taw on a temporary bridge made from scaffolding poles in approximately the same place as the above two.

A very temporary looking wooden bridge was built in 1956 (Fig. 59), also in approximately the same place as the above three, to take Belstone parishioners across the river into Skaigh Wood so that they could plant some trees there. The wood had been bought by the parish following a public appeal for funds to stop it being

59 (© Belstone Community Archive)

sold for conifer planting. Skaigh Wood had previously been part of the *Skaigh House* estate. Michael Leahy remembers that as a teenager in the Scouts he helped build the bridge using branches that had been left after contractors had felled trees in Skaigh Wood. The man crossing the rickety bridge is Cyril Robinson, the Chairman of Belstone Parish Council, who was one of the driving forces behind the purchase.

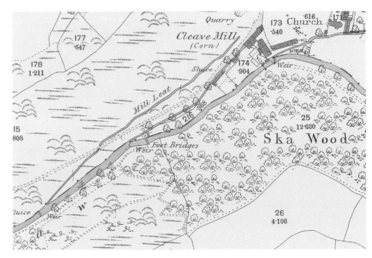

60 *Ordnance Survey 25" Devonshire LXXVII.6 1884* (extract reproduced by permission of the National Library of Scotland ©)

Two footbridges next to each other, one crossing the River Taw and one crossing the leat to Cleave Mill, are shown on the 25"Ordnance Survey map (Fig. 60). These

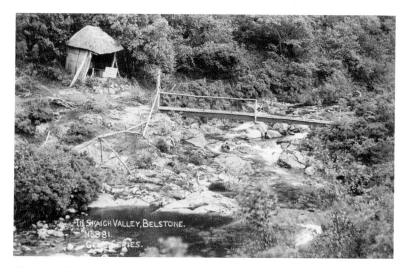

61 *In Skaigh Valley, Belstone* Gee Series postcard No 881
(© Belstone Community Archive)

were at approximately SX 637 939, not far upstream of the previous four bridges.

The three postcards (Figs. 61, 62 and 63) are of a bridge in Skaigh Wood situated where the 'Tarka Bridge' is now (SX 631 938). The wording on the sign appears to read "SKAIGH, Private Grounds, Fishing Reserved". This was a private bridge originally built for the first *Skaigh House* owner WW Symington (the house was called *Rockside* at the time). It was kept locked, presumably at

62 *Skaigh Valley bridge and summer house*, Chapman postcard 7316, postmarked 1927 (© Belstone Community Archive)

63 *Skaigh Valley*, Chapman postcard 7317 (© Belstone Community Archive)

the wood end where we can see the lattice gate. As the estate plans show, the structure next to the bridge was a summerhouse. By 1948 the bridge had become very dangerous; a public subscription raised over £33 towards the cost of renewing it, but by the 1960s it had fallen. Because the north side of the bridge was on common land the owners of Rockside/Skaigh House had to pay one shilling annually to the Lord of the Manor of Belstone.

Ivy Tor or Skaigh Bridge (SX 629 937) has been rebuilt several times since the postcard in Fig. 64 using the same stone buttresses – the following photograph (Fig. 65) was taken in May 2011, when Dartmoor National Park Rangers Ian Brooker and Pete Rich together with Paul Rendell and the local Sticklepath and Okehampton Conservation Group were rebuilding the current version. Just a few metres above the current Ivy Tor footbridge are some pieces of old iron work and a stone buttress which are the remains of another footbridge.

In 1925 a guide book author, Clinton-Baddeley, reported that "Just below Ivy Tor there is a slender bridge across the water. There is another just above, but here the hand-rail has been swept away, and the only reliable method of crossing is upon hands and knees."[45]

64 *Skaigh Bridge, River Taw*
(© Paul Rendell Collection)

65 *Ivy Tor Bridge rebuild* (Photo Chris Walpole)

66 *The Ivy Tor Mine footbridge* (© Belstone Community Archive)

67 *Ordnance Survey 25" map, Devonshire LXXVII.5 1905 edition* (extract reproduced by permission of the National Library of Scotland ©)

This view (Fig. 66) is looking west up the River Taw and Belstone Cleave to the rounded Watchet Hill, with part of Belstone Tor just visible to the left. Evidence of Ivy Tor Mine, which was active between the 1820s and 1870s, can be seen to the left and upstream of the bridge. A rifle range, whose targets were close to the mine, was established in the late 1880s (White's 1890 Directory refers to it having been 'recently made') and there is a newspaper report of men using it dated May 1891. The foundations of this bridge are still visible at SX 627 935.

A footbridge for which there is no photographic or physical evidence is shown at SX 622 935 on the 1905 25" Ordnance Survey Map (Fig. 67). This was about 200 metres downstream from the current village bridge which hadn't been built at the time of this map.

69 *Belstone Village Bridge*, Chapman postcard 7314
(© Belstone Community Archive)

68 *Belstone Village Bridge, River Taw*
(© Belstone Community Archive)

The group in this photograph (Fig. 68) at the Belstone village bridge (SX 621 933) are members of the Mortimore family who lived in the centre of Belstone. Thirteen children were born in their small cottage between 1883 and 1903. The photograph could have been taken any time between the mid-1890s through to the end of the first decade of the 1900s. Figs. 69–72 are also of the Belstone village bridge.

There were a number of other footbridges, which were probably timber constructions, further upstream

70 *Belstone Village Bridge looking towards Moorlands House*
(© Belstone Community Archive)

71

Figs. 71 and 72: *Two other views of the Belstone Village Bridge* (© Belstone Community Archive)

72

on the River Taw. A footbridge is shown on the 25" Ordnance Survey map from around 1900 at SX 622 930. This went from the land belonging to William Pike at *Moorlands House* on the Belstone side of the Taw over to the moor on the Cosdon side of the river and South Tawton parish.

A little further upstream another footbridge crossed the Taw from the private land at *Tawcroft* on the Belstone side of the river over to the Cosdon side. It is not shown on the old maps but must have been somewhere along the *Tawcroft* river frontage, i.e. approximately at SX 622 928. Both these private bridges across the river were noticed by South Tawton commoners during their Beating of the Bounds perambulation in 1914:

On the way we did several necessary pieces of business. At Dr Gwynn's [who built Tawcroft in 1908 and ran it as a Sanatorium until 1916] arrangements were made for the erection of a footbridge by him, he paying a shilling a year acknowledgment for the right to erect a base on S Tawton Common. Dr Gwynn entertained us most hospitably and showed some of us around his beautiful garden. Mr Pike [who built Moorlands House in 1891], in the same manner as Dr Gwynn, agreed to pay acknowledgement for a bridge.[46]

Finally on the Belstone section of the Taw are these references in the Parish Council minutes to a bridge which existed for an unknown length of time at Taw Marsh. Firstly on 11th April 1949 the minute reads: 'Old plank bridge in Taw Marsh had been washed away – Mr Pike to measure the river' and then on 18th May: 'Clerk to enquire price of an oak plank 24 feet long, 18" wide and 4" thick for use as a Plank Bridge beyond the first fording place in Taw Marsh'. There are no further references in the Parish Council minutes so the bridge was probably never replaced. The 'first fording place' could be a reference to a ford marked on the 6" Ordnance Survey map (1908 edition) at SX 619 915. This map also marks another ford about 100 metres further

upstream which is the current tarmac ford at Horseshoe Bend on the river at SX 620 914. The plank bridge could well have been somewhere between these two fords, but there are lots of places on this slow-moving section of the river that it might have spanned.

There has long been a footbridge over the East Okement River at Fatherford (SX 603 947), probably dating back well before the Fatherford Railway Viaduct was built just below the bridge in 1871. A footbridge is shown on the 25" Ordnance Survey map from around 1900 and there is an amusing reference to it in the Western Times of 21 September 1917 when reporting on a meeting of Okehampton Town Council:

Belstone Parish Meeting drew the attention of the Council to the dangerous condition of the footbridge at Fatherford. The bridge was erected and kept in repair by the Okehampton Council. In the course of discussion as to responsibility in the matter, Mr Hunt, who said the bridge was very much used by visitors and others attending church and coming to Okehampton, asked what would happen if, on some dark night, someone fell into the river in consequence of the dilapidated state of the bridge?

The Mayor: Ask the Coroner (laughter).

Mr Rowe: They would get a cheap bath, anyhow.

On the motion of Mr Rowe, seconded by Mr Hunt, it was agreed to consult the Parish Councils of Belstone and Okehampton as to defraying the cost of restoring the bridge, this council offering to contribute one-third of the cost.

Eddie Hain, a former Dartmoor National Park Ranger, undertook substantial repairs in 1989. The current structure, known as 'Charlotte's Bridge', is sandwiched between the railway viaduct and the A30 dual carriageway; it was opened in May 2011 after much local fundraising, ten years after twelve year old Charlotte Saunders died while crossing the swollen river on horseback.

Further upstream on the East Okement is Chapel Ford where, as the name suggests, the river was crossed by a ford with stepping stones alongside. There was probably no footbridge here until the late 1940s or 1950s. The photograph (Fig. 73) taken in 1978 shows what is believed to be the original bridge.

On the small tributary of the East Okement a footbridge is marked on the 25" Ordnance Survey map from around 1900 at SX 618 945 (Fig. 74). This is where the lane at Priestacott forded this stream (a ford is also marked on the map), the footbridge being built to save pedestrians getting their feet wet. The stream has been channelled under the lane for decades so the footbridge has long gone.

73 *Chapel Ford Bridge* (SX 608 934)
(Photo Eddie Hain © Belstone Community Archive)

74 *Ordnance Survey 25" map Devonshire LXXVII.1 1885* (extract reproduced by permission of the National Library of Scotland ©)

Two footbridges close to *Eastlake Farm*, which appear to have been only about seventy metres apart, are also marked on the same map from around 1900 at SX 612 948.

A simple modern wooden bridge takes the footpath which runs alongside the East Oakmount River over Moor Brook at SX 608 941 (Fig. 75)

75 *Bridge over Moor Brook* (photo the author)

Highdown, River Lyd

76 *High Down Ford* (© PH-L Collection)

This interesting photograph (Fig. 76) shows the ford, stepping stones and clam bridge over the River Lyd at High Down (SX 532 857) with Great Nodden in the background. Fig. 77 is a photograph of it taken after rebuilding in 1986 looking towards Great Links Tor.

Fig. 78 is a postcard of a clam bridge over the River Lyd at Wheal Mary Emma (SX 532 851). This was constructed for the inhabitants of Doe Tor Farm to access High Down in the 20th century.[47] The farmer, James Pengelly, was required in 1922, by covenant with the War Department, to keep the bridge repaired.

Field names

The most comprehensive source of historic field names comes from Tithe Maps.

Tithes were originally a tax which required one tenth of all agricultural produce to be paid annually to support the local church and clergy. After the Reformation much land passed from the Church to lay owners who inherited entitlement to receive tithes, along with the land. By the early 19th century tithe payment in kind seemed an out-of-date practice, while payment of tithes per se became unpopular, against a background of industrialisation, religious dissent and agricultural depression. The 1836 Tithe Commutation Act required tithes in kind to be converted to more convenient monetary payments called tithe rentcharge. The Tithe Survey was established to find out which areas were subject to tithes, who owned them, how much was payable and to whom. In parishes where tithes were still being paid in kind, the land had to be surveyed and valued, to arrive at total parish rent charge figures, and to calculate each individual landowner's liability to pay

77 *High Down Ford* (Photo Eddie Hain © Paul Rendell Collection)

78 *Lydford, Rivey Lyd, Showing Bray Tor* (© Stanbrook Collection)

tithe. The tithe apportionment, which was the legal document setting out landowners' individual liabilities, was accompanied by a map. These maps are a valuable source of information today.

There are sixty-three clam related field names on the Apportionments for the Tithe Maps covering Devon of which twenty-four of these are in Dartmoor parishes.[††] There are seventeen variations on the name. The most common, occurring seven times, is 'Clam Close', the second being 'Clam Park' which occurs three times. As well as the single word 'Clam', other variations on the name are 'Clam Marsh', 'Clam Close', 'Little Clam Close', 'Higher Clam Close', 'Clam Hill', 'Clam Meadow', 'Clam Bottom', 'Clam Orchard' and 'Clam Hill Coppice'.

Before claiming that all of these names were linked to the existence of clam bridges it should be mentioned that Kenneth Cameron in *English Place Names*[48] includes 'Clam Park' under a group of derogatory names for unproductive fields and John Downes, in his dictionary of Devon dialect, gives the derivation of 'cloam' as being from the Old English word 'clam' meaning 'mud' or 'clay'.[49] However, writing in 1885 William Kearley, gave the derivation of 'Clam Park' as: "The footbridge field: from 'parc', a field, and 'clam', a footbridge."[50]

Of the twenty-four, all bar five are fields that are adjacent to a stream or river and of the five none are more than two fields away and one with the name 'Clam Hill' may have overlooked a clam bridge. In one case, on the map for Hennock, the surveyor has marked a bridge on the boundary of the named field.

Not surprisingly, given the fact that clam bridges are of timber construction and depend on an easily accessible supply, all the fields are in parishes on the edge of Dartmoor or in one of the river valleys, as is the case with Widecombe. Ilsington, with five fields adjacent to a stream and one only a field away, has the most; Drewsteignton, North Bovey and Throwleigh each have three; Chagford, Hennock, Moretonhampstead and Widecombe have two; and Mary Tavy and Meavy have one each.[‡‡]

As already referred to, the word 'clam' or 'clamp' might have been applied to a field that was difficult to work and this may be the case in Throwleigh, for example, where there is a field named 'Clam[p] Close Moor' which doesn't appear to be adjacent to a stream. The presence at some point of a potato or root clamp in the field may also be the origin of the name.

An article in the *Report and Transactions of the Devonshire Association* in 1963 about field names in the parish of Widecombe referred to two fields named 'Clam Close':

There is a *Clam Close* on Jordan, on a direct line from the manor farm down the valley to Ponsworthy … For

---

[††]The boundaries of the parishes do not necessarily match those of the Dartmoor National Park and therefore this list may include fields outside the Park.

[‡‡]The clam bridges in Lustleigh are not adjacent to named fields.

the last 120 years this meadow has been known as Jordan Great Meadow, and it was only the discovery of a field list and map of 1781 that revealed that it was then known as Clam Close.[51]

At the time of the article the river was still bridged by a Clam further upstream and a modern Clam on the Two Moors Way crosses the river below Jordan Mill (Fig. 79).

79   Bridge near Jordan Mill (Photo the author)

River Bovey Clam Bridge

On a footpath from Lustleigh to Manaton below Lustleigh Cleave, what is believed to be the last surviving traditionally built clam bridge spans the River

80 *River Bovey clam bridge* (Photo Grahame Blackwell ©)

Bovey (Fig. 80) (SX 767 810). Made from tree trunks felled in the adjacent woodlands there is evidence that there has been a bridge of this construction at this site for over two hundred years:

The bridge is of the simplest character, it consists of a tree, slightly planed, and with the exception of a hand-rail on one side, may be an exact copy of the first bridge which the inhabitants of Lustleigh and Manaton used centuries ago to render more easy their journies [sic] to and fro.[52]

Shown on both the Lustleigh Tithe Map of 1837 and the 1864 Ordnance Survey map it's known that historically the tree trunks were replaced every 25 years or so. As long ago as July 1896 Lustleigh Parish Magazine reported that the tree trunks of the Clam Bridge were to be replaced because they were in a dilapidated condition. On this occasion it was repaired by a Mr. William Easton, with a tree donated by Mr Barnham. The additional cost to the Parish was £1. 15s with subscribers giving 16 shillings and the Rector making up the balance.[53]

Fig. 81 shows the bridge after repair in 1972.

There is also a ford fifty or so metres upstream of the bridge and twice there have been proposals to replace the Clam Bridge with a bridge capable of taking horses. The first proposal was made by the British Horse Society in 1991. Although initially supported by Lustleigh Parish Council, by 1993 the Council was opposing this and by May 1993 the proposal had been withdrawn.[54]

A similar proposal was made again in 2006. At this time the Clam Bridge was in need of repair not having received any attention for some years. Dartmoor National Park Authority proposed to replace the historic clam by building a bridge of sufficient width to carry horses. After strong local objections this proposal

81 *River Bovey clam bridge*
(Photo Eddie Hain © Paul Rendell Collection)

82 *The Bovey Clam Bridge alongside the modern replacement*
(Photo the author)

was dropped and, in 2007, the Authority constructed a new bridge adjacent to the existing one made from steel, wood and imported stone (Fig. 82). Following a campaign by local people that attracted national attention, Natural England took on responsibility for the Clam Bridge and it has been re-opened for use by those who prefer it to the new bridge. Its retention is a small example of the important part that artefacts like the Clam Bridge play in Dartmoor's cultural heritage.

Afterword

Around the moor there are a number of examples of wooden footbridges that probably replaced clam bridges some of which provide a less engineered response to safety. Examples include one on a public footpath over the Beadon Brook in the Teign Valley (Fig. 83) at SX 820 819.

Admittedly on private land, this bridge (Fig. 84) provides a safe crossing over the East Dart. The bridge pictured in Fig. 85 provides fishermen with access to the far side of the Cherrybrook (SX 632 738).

83 *Clam over the Beadon Brook* (Photo the author)

84 *Bridge over the East Dart near the Lydgate House Hotel* (Photo the author)

85 *Bridge at Cherrybrook Foot* (Photo Paul Rendell)

Credits and acknowledgements

Particular thanks go to Tom Greeves, formerly chair of the Dartmoor Society, who not only gave me access to many of the images in this publication, but also commented on the draft. The mass of information in the section on the bridges over the river Taw and East Okement around Belstone was provided by Chris Walpole, for which I'm deeply grateful. His research shows that there will, no doubt, be much more to learn from other parishes around the moor. I also received invaluable help from Paul Rendell.

Images also came from Grahame Blackwell, Simon Butler, Elisabeth Stanbrook; the Paul Rendell, PH-L and John Walling collections; Belstone Community Archive, British Library, Dartmoor Archive, Ivybridge Heritage and Archive Group, Lustleigh Community Archive, National Trust and Devon Archives & Local Studies Service, South West Heritage Trust.

I would also like to thank the following for additional information: Rose Cooke, Peter Davies, Demelza Hyde, Bill Murray, Martin Smith, Tanya and Barry Welch, Peter Way, Barrie Wilson and Tom Wood.

I'm grateful to the Heritage Lottery Funded Landscape Partnership Scheme, *Moor than meets the eye,* for funding this project through the Parishscapes scheme and to Emma Stockley, the Community Heritage Officer for her support.

Disclaimer

The views expressed in this publication are those of the author and do not necessarily represent the opinions of the Dartmoor National Park Authority on the subject of clam bridges on Dartmoor.

I make no claim that this survey of the history of clam bridges on Dartmoor is comprehensive and, whilst every care has been taken to ensure its accuracy, there will no doubt be errors in the text and readers will, I'm sure, be able to supply additional information. If so, this should be sent to the author at:

dartmoorclams@gmail.com

Peter F. Mason
August 2019

Notes

1. https://en.wikipedia.org/wiki/Wycoller/ Accessed 18/03/2019

2. Ordnance Survey 6" Series Devonshire XCIX SE 1885 (The clapper bridge at Bellever is also labelled 'Cyclopean') and *A Perambulation of Dartmoor* Rowe S., Plymouth 1848 p48

3. *Oxford English Dictionary* online: http://www.oed.com/ Accessed 28/01/2019

4. There are several streams on Dartmoor named Blackabrook

5. *English Dialect Dictionary* Wright J., Vol. 1 (A-C) 1898 Accessed via Internet Archive https://archive.org 28/01/19

6. *An Exploration of Dartmoor and its Antiquities,* Page JLW., London 1889 p113

7. *The Devonshire Dialect* Marten C., 1973 Exeter p16

8. *A Description of the part of Devonshire bordering on the Tamar and Tavy* Bray., Mrs 1836 Vol III pps. 264 - 266

9. *The Western Antiquary*, Vol 1, December 1881, p141

10. *A Perambulation of Dartmoor* Rowe S., Plymouth 1848 p48; quoted in *The Western Antiquary*, Vol 1, March 1882, p192

11. *The Western Antiquary*, Vol 2, April 1882, p7

12. *Worth's Dartmoor* Ed., Spooner GM., & Russell FS, Newton Abbot, 1967 edition, p370

13. *Crossing's Guide to Dartmoor* Crossing W., 1912 edition (reprinted 1965) p14

14. Information from http://www.otterton.info/Photohistory/ BridgesWeb/ottertonbridges.htm

15. Devon Heritage Centre Reference: QS/4/1748/Midsummer/ PR/30

16. *A Perambulation of Dartmoor* Rowe S., Plymouth 1848 p136

17. *A Handbook for Travellers to Devon and Cornwall* John Murray, London (6th edition, 1865) p123

18. *An Exploration of Dartmoor and its Antiquities,* Page JLW., London 1889, p113

19. *A Handbook for Travellers to Devon and Cornwall* John Murray, London (6th edition 1865) p50

20. *Observations on the Western Parts of England* Gilpin W., London 1808 pps 186/187

21. *A Perambulation of Dartmoor* Rowe S., Plymouth 1848 p221

22. Referred to by Tom Greeves in his book: *Dartmoor's Earliest Photographs, Landscape and Place, 1860 – 1880*, Twelveheads Press, 2015 p51

23. *High Dartmoor,* Hemery E., 1983 London p244

24. *Dartmoor Forest Farms* Stanbrook E., 1994 Devon Books p57

25. *Dartmoor Worker* Crossing W., 1992 edition, Peninsular Press p66

26. *Walking Dartmoor's Ancient Tracks* Hemery E., 1983 London p117

27. Information from www.legendarydartmoor.co.uk Accessed 14/01/2019

28. *Crossing's Guide to Dartmoor* Crossing W., 1912 edition (reprinted 1965) p464

29. *Folklore and Legends of Dartmoor* Crossing W., Newton Abbot 1997 p62

30. www.legendarydartmoor.co.uk Accessed 14/01/2019

31 Information from *Images of England: Dartmoor* Greeves T., Stroud 2004 p12

32 *Crossing's Guide to Dartmoor* 1912 edition; Newton Abbot 1965 p462

33 *Amid Devonia's Alps* Crossing W., London 1888 p77

34 *Home scenes: or Tavistock and its vicinity* Evans R., Tavistock 1846 p48

35 Information from *Images of England: Dartmoor* Greeves T., Stroud 2004 p121

36 *Crossing's Guide to Dartmoor* 1912 edition; Newton Abbot 1965 p426

37 Ibid., p437

38 *A Handbook for Travellers to Devon and Cornwall* John Murray, London (6th edition, 1865) p143

39 *Worth's Dartmoor* Ed., Spooner GM., & Russell FS, Newton Abbot, 1967 edition p370

40 *The Book of Meavy* Hemery P.1999, Halsgrove, p.21.

41 *Crossing's Guide to Dartmoor* 1912 edition; Newton Abbot 1965 p304

42 *Dartmoor Field Guides* Brown M., Dartmoor Press 1990s

43 Information from Paul Rendell

44 *Crossing's Guide to Dartmoor* 1912 edition; Newton Abbot 1965, p430/431

45 *Devon* Clinton-Baddeley VC., London 1925

46 *The Book of Belstone* Walpole C. & M., Okehampton, 2002

47 Doe Tor Farm, Lydford, Devon Greeves, T., 2017 (Wessex Archaeology, Salisbury 62760.01) p.10

48 *English Place Names* Cameron K., London 1963 p209

49 *A Dictionary of Devon Dialect* Downes J., Padstow 1986 p73

50 *The Western Antiquary* Vol 4, March 1885 p203

51 *Field Names in Widecombe-in-the*-Moor, (1963) H. French, *Report and Transactions of the Devonshire Association,* Vol 95 p162

52 *Lustleigh Parish Magazine* June 1904

53 *Lustleigh Parish Magazine* July, August, October 1896

54 Lustleigh Parish Council minutes 14/5/1991 and 6/4/1993